RAPID Manufacturing Institute for Process Intensification 2022

Held at the 2022 AIChE Annual Meeting

Phoenix, Arizona, USA
13-18 November 2022

ISBN: 978-1-7138-7922-0

Printed from e-media with permission by:

Curran Associates, Inc.
57 Morehouse Lane
Red Hook, NY 12571

AIChE
120 Wall Street, FL 23
New York, NY 10005-4020

Phone: (800) 242-4363
Fax: (203) 775-5177

www.aiche.org

Additional copies of this publication are available from:

Curran Associates, Inc.
57 Morehouse Lane
Red Hook, NY 12571 USA
Phone: 845-758-0400
Fax: 845-758-2633
Email: curran@proceedings.com
Web: www.proceedings.com

TABLE OF CONTENTS

RAPID POSTER SESSION

732a Spatial and Temporal Intensification of Poultry Processing Wastewater Treatment Units Using Stainless Steel Ultrafiltration Membrane

Saubana Dada, Mahmood Gheni Jebur, Ranil Wickramasinghe

The current treatment of poultry processing wastewater (PPW) requires a large expanse of land, the use of toxic chemicals, and takes time. The wastewater is typically treated prior to discharge. Apart from aiming to reuse the treated water for non-food activities, this project aimed to reduce the footprint and time required for PPW treatment. To intensify the PPW treatment units, we studied the possibility of replacing dissolved air floatation (DAF) with a stainless steel ultrafiltration membrane (SSUF). Combined PPW from all processing units taken before the first DAF was used for this study with no pretreatment. The SSUF used for this study has a pore size of 0.02 μm, and the performance of the SSUF membrane was studied by measuring the flux at 40 psi, 70 psi, and 110 psi transmembrane pressure. The flux was normalized to 250C.

The particle size analysis was carried out to determine the distribution of particles in the PPW. A cleaning procedure that entails using alkali and acid was developed for the SSUF. COD, BOD, TSS, TKN, PH, oil, and grease were measured before and after each experiment. The result shows that the flux became steady at 30 L/m2h after 2 hours of experiment, irrespective of the TMP. The SSUF membrane removed TSS 99.9%, oil and grease 99.9%, COD 90%, BOD 90%, nitrogen 76%, and soluble BOD 60%. The removal efficiency was higher at 110 psi. The result obtained was compared with the data from the industry. Finally, the membrane removed E. coli and coliform up to 99.9%, which validated the pathogen removal ability of the SSUF. In conclusion, the results show that SSUF achieved comparable performance to that of the current treatment used for the PPW treatment.

732b Intensified CO_2-to-Ethylene Process in Electrochemical Cells at Elevated Temperatures (350-500 °C)

Lucun Wang, Min Wang, Yingchao Yang, Dong Ding

Ethylene is a primary building block for the organic chemical and polymer sector and carbon negative ethylene could provide a pathway to decarbonize the large volume products including plastics and even sustainable aviation fuels (SAF). In this work, a carbon negative electrochemical process is developed for direct conversion of CO_2 to ethylene at intermediate temperatures (350-500°C). The new process is based on high-performance protonic ceramic electrochemical cells (PCECs) that integrate CO_2 reduction reaction to produce ethylene in the cathode and concurrent H_2O oxidation reaction in the anode to provide hydrogen for CO_2 conversion. Highly active and selective catalysts for CO_2 to ethylene conversion have been identified and integrated into the PCECs. Both the catalytic and electrochemical performances of the integrated reactor system are evaluated and optimized.

732c Thermodynamic Modeling of Double Azeotrope - a Study of Methanol-Diethylamine Binary Mixture

Nooram Anjum, Shigeo Oba, Toshihiko Hiaki, Chau-Chyun Chen

'Azeotrope is a mixture of liquids which boil at a constant temperature at a given pressure. Although such mixtures are difficult to separate, they have proven useful to overcome various challenges such as minimization and recovery of waste of chemical processes, thus helping build a greener world. Therefore, it renders development of a thermodynamic model to correlate vapor liquid equilibrium conditions for mixtures exhibiting such behavior. In general, one stable azeotrope formation occurs for a mixture for a given VLE conditions. The property of a mixture that produces more than one stable azeotrope at a given temperature or pressure is known as polyazeotropy. Representing the VLE data for a mixture exhibiting polyazeotropy has been of interest for many years.

The present study focuses on the binary mixture consisting of Methanol and Diethylamine. This mixture involves self-association between the components as well as cross association which is why mixed deviations from Raoult's law are observed. This behavior is not observed at low pressure for a constant temperature, but as the pressure increases double azeotropy can be detected. This study includes the analysis of association extents (bounded or non-bounded sites) to explain how the double azeotropes are related to the associations. Correlating the VLE data using UNIQUAC/UNIFAC model is insufficient as it only works for low and moderate pressures. Since the system exhibits cross and self-association, Nonrandom two-liquid (NRTL) model is also not well suited as it is does not account for specific chemical associations. Therefore, in order to represent the given system, Non-random two liquid activity coefficient model with association theory is used. This model combines Wertheim's perturbation theory for associations between the constituents with the classical NRTL model accounting for the non-idealities in the liquid phase. It regresses binary interaction parameters and unary molecule specific parameters to provide a satisfactory correlation for the experimental VLE data obtained from literature and successfully correlate the VLE data for the given binary mixture for a wide range of pressures.'

732d Enhanced Liquid-Liquid Separation Using a 3D-Printed 2-Dimensional Device with Advanced Coatings

Lei Li, Aigerim Baimoldina, Fan Yang, Yihan Song, Cliff Kowall, Glenn Cormack

Wastewater treatment remained critical for the chemical industry in general. Application and adoption of intensified process design and 3D-printed devices offer the prospect of revolutionizing the oil-water separation. A 3D-printed device that takes the advantage of potential reduced size, increased scalability of equipment, and process intensification through faster extraction and phase separation at a lower energy cost could be highly amenable to modular chemical manufacturing. To this interest, our collaborators from Oregon State University designed a 3D-printable liquid-liquid separation device for oil-water phase separation via guiding the mixture flow through a capillary force gradient 'field'. The operation is designed to not allow a selected phase to overcome capillary interface forces with inertial and viscous forces, thus guiding the fluid towards only one outlet stream. To further improve the separation efficiency, we have designed and fabricated the coatings with the desired wettability gradient, i.e., from hydrophilicity/oleophobicity at one end to hydrophobicity/oleophilicity at the other end. Simultaneous hydrophilicity/oleophobicity has been achieved with Perfluoropolyethers (PFPEs) coating on glass slides. A lower water contact angle (WCA) of $< 30°$ and a higher hexadecane contact angle (HCA) of $> 65°$ were obtained. In addition, such coating is durable after 100-hour immersion in water and organic solvents, evidenced by the unchanged HCA and coating thickness. Meanwhile, simultaneously hydrophobic/oleophilic surface coating has been achieved with chloro(dodecyl)dimethylsilane-based SAMs, where WCA is 90° and HCA is below 10°. This coating is also stable after 100-hour immersion in water and organic solvents. Our work demonstrates the potential of the advanced coatings in application of liquid-liquid separation.

732e Efficient Chemicals Production Via Chemical Looping

Sunkyu Kim, Jian Pan, Arun Senthil Sundaramoorthy, Raul Lobo

This project develops the chemical looping (CL) technology into a general process intensification for natural gas upgrading to commodity chemicals. Nonoxidative upgrading of methane dehydroaromatization (MDA) to aromatics is often limited by thermodynamics, which could be alleviated by CL from selective oxidation of H2 product and H2O adsorption on the molecular sieve. Silica-encapsulated Fe2O3 is implemented for the CL, and a silica shell can inhibit larger molecules of aromatics from diffusing to the Fe2O3 surface, enhancing overall MDA performance by selective H2 combustion. The developed system provides high methane conversion and aromatic yield of ~50% and ~35%, respectively, in a recirculating mode for 4 hours. In addition, we develop an optimization and control relevant dynamic model for the modular integrated process, consisting of MDA and CL by using a restricted set of process information. A novel iterative modelling framework integrating active learning principles with state estimation techniques is proposed. The proposed approach will allow us to learn the integrated process model iteratively through experimentation even when the process data is limited. For further development of other natural gases (ethane or propane), In, Ga, or Mn-based zeolites with different Si/Al and metal/Al ratios are synthesized and used to catalyze alkane dehydrogenation. A novel type of Mn supported on zeolite are found to have high reaction activity and high selectivity. The reaction process and potential working mechanism are investigated using in-situ Raman and diffuse reflectance infrared Fourier transform spectroscopy (DRIFTS).

732f Development and Demonstration of Novel Thermal Technologies for Enhanced Air - Side and Two - Phase Performance of Cpi - Relevant Heat Exchangers

Matthew Realff, Ari Glezer, Srinivas Garimella, Arne Pearlstein, Thomas Lestina

'Almost every process in the chemical process industries (CPI) involves heat transfer. Georgia Tech (GT), the University of Illinois at Urbana Champaign (UIUC), and Heat Transfer Research Inc. (HTRI) are working on an integrated RAPID project to develop, test, and demonstrate, in a testbed industrial condenser, novel air- and process-side heat transfer technologies developed at the laboratory scale at GT. The goal is to improve air- and process-side thermal performance, by using autonomously fluttering reeds on the air side, and ultrasound transducers on the process side, to significantly improve energy utilization, and reduce footprint, weight, and cost. These technologies help overcome limitations of previous passive, geometrically complex designs burdened with significant flow losses, and can be incorporated into new and existing equipment.

Laboratory work at GT has focused on demonstrating 20-30% (air- and process-side) enhancement. On the air side, GT has explored reed geometries, materials, and array configurations leading to two-fold heat transfer enhancement with minimal pressure drop penalty. On the process side, GT has investigated modulation frequency, power, and array configurations for ultrasound transducers leading to complete condensation, thereby allowing significantly reduced condenser length, mass, and volume. We have addressed challenges associated with durability/insertion of reed arrays and ultrasound transducer placement. Air-side work at UIUC has pointed the way to use of a streamwise series of reeds and has developed computational tools to scale the GT laboratory results to pilot-scale demonstration at HTRI. Process-side models account for interface shape, bubbles in the liquid phase, and drops in the vapor phase. Work at HTRI incorporates the reed and ultrasound technologies into a pilot-scale condenser, and measurements are under way to quantify performance under real-world conditions, including air- side flow maldistribution. Process economics work at GT accounts for capital and utility costs, and provides upper bounds on fabrication and installation costs below which the new technologies are economically viable. This work shows how the new technologies can increase throughput when improved heat transfer de-bottlenecks a process, and how one can eliminate equipment that would be needed for recirculation and phase disengagement of incompletely condensed fluid when improved condenser performance so allows.

These novel heat transfer approaches have the potential to transformatively enhance performance of a broad class of heat exchangers across many CPI applications. They also have the potential to enable new applications, including natural gas upgrading by cryogenic distillation, multifunctional modules, and utilization of renewable bioproducts, as well as integration of multiple processes (e.g., separation and heat transfer for binary and multi-component liquids, for which ultrasound has demonstrated the capability to produce an atomized mist enriched in one component). Higher energy efficiency and energy productivity allow for new trade-offs between reduced footprint/mass (and hence lower capital cost), and higher throughput (allowing for de-bottlenecking). The potential for swap-in/swap-out retrofit is particularly attractive.'

732g Microwave Catalysis for Process Intensified Modular Production of Value-Added Chemicals from Natural Gas

Brandon Robinson, Ashley Caiola, Yifan Deng, Sanjana Karpe, Dushyant Shekhawat, Joseph Powell

'The microwave-assisted catalytic conversion of methane to value-added chemicals was explored over a 4-wt% Mo/ZSM-5 catalyst premixed with 40-wt% SiC powder. A benzene yield of 3.5% was achieved at a catalyst surface temperature of 550°C. Regeneration experiments were performed under 2% O2/N2 under microwave at 100 W/500°C. Benzene yields dropped after the first regeneration cycle but stayed constant for the subsequent two cycles. The design of an H-Field cavity increased the testing capacity to a maximum of 20g of catalyst. A linear increase in yield with the increase in catalyst mass suggests that at the current rate of aromatics + C2 production could be produced at 2043 g/day.kg-1catalyst. Process simulation and technoeconomic analysis results indicate that energy efficiency improvement of 58% can be achieved compared to the base case. The energy penalty for the new MW-assisted catalysis process is about 2x times less than the base case. The technoeconomic analysis resulted in a positive net present value for the microwave-assisted plant-wide model.

To further increase the efficiency of the microwave, a microwave-sensitive SiC@Fe-ZSM-5 core@shell catalyst was designed to enhance the direct heating of the zeolite. It was found that the Core@ Shell design increased the heat transfer to the Fe/ZSM-5 more efficiently than the physically mixed SiC catalyst. The temperature of the zeolite phase in the core@shell catalyst was higher than that of the physically mixed catalyst, thus resulting in a higher C6 yield. A temperature-indicating probe molecule was used to estimate the catalyst's temperature distribution and metal site temperature under microwave heating. The scaled-up production of the microwave-sensitive Fe-ZSM-5 catalyst was also explored through the design of a continuous zeolite synthesis system. The designed synthesis system yielded a 100x increase in the space-time yield of zeolite produced compared to lab-scale batch processing of six 50-mL batches per 6 days.
'

732h Integrating Reaction and Separation in a Thin Film Evaporator for Intensified Production of Dispersants

Riddhesh Patel, Nasser Al Azri, Hari Mantripragada, Nicolas Proust, Glenn Cormack, Robert M. Enick, Goetz Veser

'Specialty chemicals are to-date largely produced in large batch reactors with typical volumes of tens to hundreds of thousands of gallons. While these batch processes are simple to operate, they are also highly inefficient. We previously demonstrated that transition from batch to continuous production can result in strong improvements in energy efficiency as well as reductions in physical footprint and capital cost. The investigations used the production of succinimide dispersants as example process. These dispersants are produced by a two-step reaction between poly-isobutylene succinic anhydride (PIBSA) and polymeric amines. The first step is an addition reaction to form a succinic amide intermediate, followed by a reversible dehydration step to produce the desired polyisobutylene succinimide product and water as byproduct. In the conventional batch process, the produced water evaporates into the head space of the reactor from where it is constantly removed via a purge stream. In contrast, in the continuous reactor water remains in the reacting stream, resulting in equilibrium limitations and necessitating a downstream drying step to obtain the dehydrated dispersant product.

We had previously identified an agitated thin film evaporator (TFE) as well-suited for drying of the highly viscous product stream. Here, we are investigating the use of the TFE as a reactive separator, i.e., using the TFE as an intensified unit in which reaction and separation occur concurrently. The TFE was first modeled by integrating previously developed reaction kinetics with mass and energy balance equations. Appropriate correlations for heat and mass transfer were identified from the literature and experimentally verified for our set-up. This performance model was then used to probe the feasibility of replacing the two-step reactor-evaporator system with a single TFE as an integrated reactive separator. The model predictions were validated experimentally in the final step. The results from our study confirm that a thin film evaporator could indeed serve as a highly intensified, integrated reactor-separator unit which is particularly well-suited for reactive processing of viscous mixtures in the presence of equilibrium limitations. '

732i Modelling for Optimal Operation of Modular Integrated Methane Dehydroaromatization Process

Arun Senthil Sundaramoorthy, Sunkyu Kim, Babatunde A. Ogunnaike, Raul Lobo

'Natural gas is a viable alternative to crude oil-based feedstock to meet energy and chemical manufacturing demands. However, many of these deposits are in remote locations, motivating the development of on-site manufacturing approaches, known as modular technologies, for effective process intensification and economical production of desired products [1, 2]. Herein, a novel modular methane dehydroaromatization process integrating dehydroaromatization (DHA), chemical looping (CL), and temperature swing adsorption (TSA), proposed by Brady et.al [3] is considered. Experiments have proven the process to be a promising technology in terms of yield and energy consumption compared to conventional process technologies [4, 5]. However, the dynamics of the modular process operated in semi-batch units, involving interactions due to material re-circulation and non-linear rate kinetics, are yet to be understood from an operations perspective. Understanding the process operation is challenging because process variables are not easily measured, limiting the amount of process data needed to verify the process of interest.

This work aims to develop a dynamic model for the modular integrated dehydroaromatization process that can describe the operating space by elucidating the process dynamics using a restricted set of process information. A novel iterative modeling framework is proposed that integrates active learning principles with state estimation techniques [6, 7, 8]. The iterative modeling framework consists of an integrated model development step for optimal dynamic experiment design and a state estimator to estimate the restricted information in the process of interest [9, 10]. This approach will allow us to learn the integrated process model iteratively through experimentation, even when the availability of process data is restricted. These models will be used to develop safe and optimal process operation strategies.

As a first step, an integrated model is being developed for the process of interest to design optimal experiments and develop a state estimator. A stand-alone process model developed for DHA, CL, and TSA, as advection-diffusion-reaction partial differential equations, were numerically simulated using the method of lines. The verified stand-alone process models are being integrated, considering material re-circulation in semi-batch operation, using a simultaneous modeling approach incorporating a singular perturbation technique to obtain the solution [9]. Later, state estimators will be developed to extract information and update the integrated process model to design the experiments.

References

[1] Baldea Michael, Edgar Thomas F, Stanley Bill L, Kiss Anton A. Modular manufacturing processes:
Status, challenges, and opportunities AIChE journal. 2017;63:4262-4272.

[2] Bielenberg James, Palou-Rivera Ignasi. The RAPID Manufacturing Institute - Reenergizing US efforts
in process intensification and modular chemical processing Chemical Engineering and Processing. 2019;138.

[3] Brady Casper, Debruyne Quinten, Majumder Ankita, et al. An integrated methane

dehydroaromatization and chemical looping process Chemical Engineering Journal. 2021;406:127168.

[4] Zichittella Guido, P´erez-Ram´ırez Javier. Status and prospects of the decentralised valorisation of natural gas into energy and energy carriers Chemical Society Reviews. 2021;50:2984-3012.

[5] Ren Tao, Patel Martin, Blok Kornelis. Olefins from conventional and heavy feedstocks: Energy use in
steam cracking and alternative processes Energy. 2006;31:425-451.

[6] Bonvin D, Georgakis C, Pantelides CC, et al. Linking models and experiments Industrial andamp; Engineering
Chemistry Research. 2016;55:6891-6903.

[7] Settles Burr. Active learning Synthesis lectures on artificial intelligence and machine learning. 2012;6:1-114.

[8] Dochain Denis. State and parameter estimation in chemical and biochemical processes: a tutorial Journal of process control. 2003;13:801-818.

[9] Baldea Michael. From process integration to process intensification Computers andamp; Chemical Engineer-
ing. 2015;81:104-114.

[10] Franceschini Gaia, Macchietto Sandro. Model-based design of experiments for parameter precision:
State of the art Chemical Engineering Science. 2008;63:4846-4872.

732j Process Modeling and Simulation of Electrochemical Processes Poster

Sadia Saberin, Chau-Chyun Chen

'Abstract
Numerous electrochemical technologies are being developed for energy and sustainability. Examples include electrochemical processes for chemical synthesis, batteries for power storage, fuel cells for power generation, electric vehicles, semiconductors, waste treatment, water purification, biosensors, etc. With higher efficiency and less environmental footprint, electrochemical technologies garner heightened interest among academia and industries to develop and optimize these processes. Yet, the knowledge of electrochemical fundamentals is limited to specialists and excluded from the model library of chemical process simulators readily accessible to millions of chemical engineers [Ref 1]. To accelerate the application of electrochemical technologies in the chemical industry, it is important to capture the knowledge of electrochemical fundamentals in process simulators to encourage chemical engineers to practice electrochemical engineering and to expand the scope of this promising field. This paper presents a concise summary of electrochemical fundamentals essential to model and simulate electrochemical processes. The electrochemical fundamentals include basic concepts in electrochemistry, transport phenomena, and thermodynamics. The paper also covers a review of state-of-the-art in modeling electrochemical processes and presents an example in modeling and simulating a commercially important electrochemical process to produce adiponitrile from acrylonitrile. Furthermore, this paper suggests development of specific model library modules in process simulators for electrochemical engineering.

References

1. C.-C. Chen, M.B. Gorensek, J.F. Scott, I. Palou-Rivera, "Challenges and Opportunities in Process Modeling of Intensified Chemical Processes," Chemical Engineering Progress, 2020, 116, 35-42.
'

732k Argonne Capabilities in Chemical Process Intensification

Aaron Fluitt, Meltem Urgun-Demirtas

This poster will outline Argonne National Laboratory's capabilities that are relevant for chemical process intensification, including process development and scale-up, process simulation and modeling, life-cycle analysis and techno-economic assessment, advanced materials, characterization, and AI/ML.

732l Separation of H2 Using Fluorinated Polymer Membranes RAPID Project 6.9

Abby Harders, Micah Welsch, Rajkumar Kore, Brian Laird, Kyle Camarda, Alan Allgeier, Mark B. Shiflett

Membrane technology for the separation of H2 is a widely sought-after technology. Developing high performance, economically feasible polymeric membranes for this gas separation is of major interest to the H2 purification industry. The use of fluoropolymers for H2 separation provides the opportunity to achieve a high selectivity separation while maintaining high flux. RAPID Project 6.9 is investigating the use of perfluoro(butenyl vinyl ether) and perfluoro(2,2-dimethyl-1,3-dioxole) for the separation of H2 from N2 and CO2. Experimental permeability measurements have been obtained using the pressure-rise method. Molecular simulation and structural predictive models are being utilized to provide insight on synthesis and understanding fundamental solubility and diffusivity behavior of gases in fluorinated polymers. Promising materials are being developed into hollow-fiber membranes in order to measure mixed gas selectivity and throughput to meet the industrial challenge of this separation.

732m Optimizing Absorption to Improve Haber-Bosch Synthesis Poster

Alon McCormick, Cory Marquart, Chinomso Onuoha, Matthew Palys, Zac Pursell, Edward L Cussler, Prodromos Daoutidis, Paul Dauenhauer, Michael Reese, Sameer Parvathikar, Mahdi Malmali

'Ammonia absorber columns offer an alternative separation unit to replace condensation in the Haber-Bosch synthesis loop. Metal halide salts can selectively separate ammonia from the reactor outlet gas mixture and incorporate it into their crystal lattice with remarkably high thermodynamic capacity. In our previous work, we had found that while the salt working capacity can be limited and unstable when in its pure form, the working capacity can be stabilized and increased by formulating the absorbent with an inert support.

In this RAPID project we undertook to find and understand optimal conditions for uptake and release of ammonia, guided in large part by technoeconomic analysis that takes into account the experimental data acquired. The production capacity (ammonia processed per unit absorbent and per unit production time) depends on processing parameters - including uptake and regeneration temperatures, ammonia partial pressures, release time, and sweep gas flow rate. Process simulation using the information from experiment helped to inform about optimization needs, which were governed by technoeconomic analysis. As just one example, we found that it is necessary to balance the release time with the full cycle time and bed size to ensure that the uptake breakthrough time makes efficient use of the bed. These parameters are mutually interdependent; their optimization can be nontrivial but rewarding.

The analysis shows that at the production rate for the benchmark design (30 metric tons per day), that optimized absorber-enhanced design provided lower levelized dost of ammonia synthesis and lower energy demand in comparison to optimized condenser-based design - meeting RAPID targets. As the production scale decreases (toward more distributed, small-scale production facilities) the advantage of the absorber-enhanced design increases. The results suggest that absorber-enhanced design may help make distributed ammonia production viable in a wider range of circumstances.
'

732n Unlocking the Power of Sewage Sludge and Biosolids through the Integration of Supercritical Water Technologies to Produce Renewable Natural Gas

David Kenney, Geoffrey Tompsett, Michael R. Thorson, Jesse Bond, Michael T. Timko, Ignasi Palou-Rivera, Andrew R Teixeira

'WPI has partnered with RAPID and several national labs on two recent DOE awards reimagining waste to energy routes by hydrothermal liquefaction of wet wastes. While the university teams are discovering new chemical pathways, RAPID is evaluating the technoeconomic feasibility and identifying viable pathways to market transformation.

The United States generates approximately 34 billion gallons of wastewater every day. Current wastewater treatment (WWT) facilities meet the rate of waste generation but are energy intensive, consuming nearly 30 TW-h annually. WWT generates undesired sewage sludge and biosolid slurries that are often discarded to landfills; however, these waste streams typically have an energy content 5-times the energy required to operate a WWT facility.

WPI and RAPID are collaborating to effectively manage the sewage sludge waste stream by converting it to a renewable natural gas (RNG). Hydrothermal gasification (HTG) is a method proven to convert organic waste feeds into RNG. At WPI, we are looking to integrate additional supercritical water technologies such as hydrothermal liquefaction (HTL), supercritical salt precipitation (SCSP) to eliminate issues associated with solid and inorganic loading on the HTG reactor. Rapid will then take the results shown in our studies to build comprehensive technoeconomic and life cycle analyses that will help justify launching this technology in today's market'

732o Deducing Subnanometer Heterogeneous Supported Catalyst Structure from Infrared Spectroscopy

Vinson Liao, Maximilian Cohen, Yifan Wang, Dionisios Vlachos

'Adsorbate vibrational excitations, probed via infrared (IR) spectroscopy, are sensitive to adsorbate/metal interactions, accurate, and easily obtainable in-situ or operando. Most IR-based peaks, however, are typically assigned heuristically for relatively simple spectra following the gold standard of well-defined single crystals. Deducing the structure of highly dispersed heterogeneous catalysts consisting of single-atoms and ultra-small clusters directly from IR spectra has been lacking and is imperative for intelligent catalyst design for structure sensitive chemistries.

Here, we present a computational framework to characterize supported single-atoms and subnanometer clusters from adsorbate vibrational excitations determined from IR spectroscopy. We bypass the vast combinatorial space of brute-force spectra matching by determining a viable, low-energy ensemble of structures using machine-learned Hamiltonians, genetic algorithm optimization, and grand canonical Monte Carlo calculations that contributes maximally to the final spectral intensity. We obtain first-principles vibrations on this tractable ensemble and generate single-cluster primary spectra analogous to pure component gas-phase IR spectra. These spectra act as calibration standards that allow us to predict cluster size and shape distributions, and quantify uncertainty, within the Bayesian Inference (BI) framework directly from experimental and computational data. We demonstrate the efficacy of our BI deconvolution methodology on a case study of CO adsorption on Pd/CeO2(111) catalysts and find that our conclusions are consistent with those made from other characterization techniques. Finally, we discuss extensions for characterizing complex materials towards closing the materials gap. '

732p Facilitating Batch to Flow Process Conversion through Hands-on Training on Heat and Mass Transfer Characterization, Multiphasic Systems, and Microfluidic Equipment

Fatou Baka Diop, Ashli Silvera, Gabriela Chong, Andrew R Teixeira

The paradigm shift undertaken by the pharmaceutical and specialty chemical industries from batch to continuous-flow processing necessitates a strong understanding of the benefits of continuous operation which can be more controlled, cost-effective, and efficient than traditional batch. The work presented herein is a RAPID-funded course development project that strategically combines industry and academic expertise to provide hands-on training on key principles of batch to flow process conversion with case studies on heat transfer, mass transfer, multiphasic systems, and process automation. The training is divided into four modules, each centered around a case study relevant to the pharmaceutical industry teaching fundamentals of translating a process from batch flow and its advantages on throughput, safety, and product quality. The first module focuses on translating the exothermic decomposition of hydrogen peroxide from a batch reactor where a thermal runaway occurs, to a flow reactor that enables high temperature and pressure operation. Moreover, heat transfer coefficients and residence time distributions were obtained for batch and flow reactors to model temperature and concentration profiles and choose optimal reaction conditions for high throughput and safe operation. In flow, applying 100psi back pressure to compress the oxygen produced and increase liquid residence time led to full conversion at 70oC for a theoretical residence time of 10 minutes. The second module quantifies micro-mixing in batch and flow systems using the Villermaux-Dushman method for fast single-phase reactions in which segregation results in byproduct formation. The reactive liquid-liquid extraction of salicylic acid with integrated membrane separation is used to compare batch and flow performance and highlight the use of different micromixers in flow. The flow system with static mixer reached full conversion in a minute with a mass transfer coefficient two orders of magnitude higher than in batch where 70% was reached in an hour. The third module focuses on packed bed reactors, specifically pressure drop as a function of packing particle size and residence time distributions to quantify the degree of mixing. The curriculum developed in this work has been used to train industry professionals, professors, and graduate students in the pilot RAPID course. It was also offered to a group of high school and undergraduate women, showing the adaptability of the material to technically-diverse audiences.

732q Physical Property Data and Models in Support of Bioprocessing Separation Technologies for Organic Acids Separation

Meltem Urgun-Demirtas, Ignasi Palou-Rivera, Chau-Chyun Chen, Lauren Valentino, Simon Leyland, Jean Yves Delannoy

This project forms an alliance of academia-industry-national lab partners to develop a methodology for modeling the full process of bio-separation units for both single and multi-component systems. The methodology includes the collection and incorporation of physical property data into advanced thermodynamic models, the development of advanced thermodynamic models, and their connection to commercial process simulation environments, supporting the scale-up and development of membrane and (electro)sorption-based processes for separation and recovery of organic acids from aqueous solutions (e.g., fermentation broth, lignin-rich streams, etc.). By bringing together a combination of experimental and modeling expertise, the project aims to facilitate the scale-up predictions and design of key bioseparation technologies for organic acids separations. The project team leverages a systematic experimental data generation to build thermodynamic models that will be input into bioseparation process models along with the relevant kinetic parameters. The process models will be compared and verified to scale up separations processes for pilot experiments. The simulation software vendor partners will support and guide the implementation and application of the data and models in commercial process simulators. The development of advanced Machine Learning (ML)-based models to speed up the simulation performance will also be considered.

732r Modeling the Distribution of Supported Sub-Nanometer Cluster Catalysts Poster

Salman A. Khan, Stavros Caratzoulas, Dionisios Vlachos

'Supported sub-nanometer clusters are active and selective for various industrially important reactions.1 Unlike ordered catalysts, like metal/metal oxide crystals and zeolites, supported clusters present a distribution of active sites and they cannot be precisely characterized using spectroscopic methods. Furthermore, they present an intractably large configurational space, which cannot be reliably sampled using expensive ab initio computational methods. To develop a fundamental understanding of these catalysts, it is critical to predict the structural distribution of clusters under operation conditions. Here we extend our recent work from predicting the structure of monometallic clusters to bimetallic clusters.2 We find that bimetallics are inherently more complex and do not necessarily reside on a lattice but reasonable approximations could be made. We develop a computational framework to model the distribution of supported clusters using surrogate machine learning models. We apply the method to predict the distribution of PtSn clusters supported on oxide supports and predict the structure of low energy clusters and the effect of composition. Our framework can be readily applied to other supported cluster catalysts.

References

1. L. Liu, and A. Corma, Metal Catalysts for Heterogeneous Catalysis: From Single Atoms to Nanoclusters and Nanoparticles. Chem. Rev. 118(10), 4981-5079 (2018). DOI:10.1021/acs.chemrev.7b00776.

2. Y. Wang, Y.-Q. Su, E. J. M. Hensen, and D. G. Vlachos, Finite-Temperature Structures of Supported Subnanometer Catalysts Inferred via Statistical Learning and Genetic Algorithm-Based Optimization. ACS Nano 14(10), 13995-14007 (2020). DOI:10.1021/acsnano.0c06472.

'

732s Integrated Electrocoagulation/Ultrafiltration-Membrane Distillation-Crystallization for Treating Hydraulic Fracturing Produced Water

Ranil Wickramasinghe, Mahmood Gheni Jebur, Yelyzaveta Bachynska

'In this work, produced water (PW) generated from hydraulic fracturing was treated using an integrated electrocoagulation/ultrafiltration, membrane distillation and crystallization processes (EC/UF-MDC). The focus of this work was to determine the feasibility of this integrated process for increasing water recovery. The results of this work suggest that optimizing the various unit operations in this integrated process could be used to recover PW.
All membrane based separation processes suffer from membrane fouling. Pretreatment of the feed is essential to suppress fouling of the membrane. Here electrocoagulation (EC) followed by ultrafiltration (UF) was used to achieve high removal efficiency of both total suspended solids (TSS) and total organic carbon (TOC). Dissolved organic compounds are known to foul the hydrophobic membrane used in MD. In this study, a significant reduction in membrane fouling was obtained, which can lead to a long-term durability of MD system. In addition, the use of membrane distillation crystallization (MDC) can help mitigate the scale formation. Crystallization in the feed tank was used to suppress scale formation on the MD membrane. The integrated EC/UF-MDC process can have a potential impact on Water Resources/Oil andamp; Gas Companies. By treating and reusing PW, preservation of surface and groundwater forming 80% of the water utilized in hydraulic fracturing could be achieved. In addition, treating PW will reduce the amount of PW directly disposed in Class II disposal wells, which further address the main cause of earthquakes.
'

732t Towards a Software Prototype for Synthesis of Operable Process Intensification Systems

Moustafa Ali, Navya Pabba, Shivam Vedant, Dustin Kenefake, Efstratios N. Pistikopoulos

'Process Intensification (PI) is an extensively used strategy to improve the sustainability and energy efficiency of chemical processes while also lowering the overall cost of production (Moulijn et al., 2008). There has been a burgeoning interest in PI among the process systems community as evidenced through the development of new technologies and methodologies in academia and industry alike (Tian et al., 2018). However, research towards creating software tools to leverage the state-of-the-art tools is still lacking. Driven by the recent national initiatives, such as RAPID Institute (Bielenberg and Palou-Rivera, 2019), several projects have been actuated to offer computer-aided development and analysis of PI systems (Tula et al. (2017); Demirel et al. (2017); Miller et al. (2018)).

In this work, we present an initial attempt towards a software prototype housing bespoke suites for holistic and targeted development of PI flowsheets (Vedant et al., 2021). The platform consists of three major suites: (i) Synthesis Suite - to enable discovery of novel and intensified processes, without prepostulation of potential unit operations or equipment, through phenomenon-based flowsheet generation (Papalexandri and Pistikopoulos, 1996); (ii) Model Library - to house numerous models that can be implemented within the prototype for synthesis, validation, and operability analysis of PI systems; (iii) Operability Suite - to enable dynamic optimization of PI systems and provide advanced model-predictive control strategies for online implementation (Pistikopoulos et al.,2015). A Python-based user interface (UI) is developed to seamlessly implement the integrated generation of PI flowsheets. Furthermore, a case study on pentene metathesis is presented to demonstrate the capabilities of the platform for deploying safe and operable PI systems.

Keywords: Process Intensification, Process Synthesis, Process Operability, Software Prototype

References

J. Bielenberg, I. Palou-Rivera, 2019. The rapid manufacturing institute-reenergizing us efforts in process intensification and modular chemical processing. Chemical Engineering and Processing-Process Intensification 138, 49-54.

S. E. Demirel, J. Li, M. F. Hasan, 2017. Systematic process intensification using building blocks. Computers andamp; Chemical Engineering 105, 2-38.

D. C. Miller, J. D. Siirola, D. Agarwal, A. P. Burgard, A. Lee, J. C. Eslick, B. Nicholson, C. Laird, L. T. Biegler, D. Bhattacharyya, et al., 2018. Next generation multi-scale process systems engineering framework. In: Computer Aided Chemical Engineering. Vol. 44. Elsevier, pp. 2209-2214.

J. A. Moulijn, A. Stankiewicz, J. Grievink, A. G´orak, 2008. Process intensification and process systems engineering: a friendly symbiosis. Computers andamp; Chemical Engineering 32 (1-2), 3-11.

K. P. Papalexandri, E. N. Pistikopoulos, 1996. Generalized modular representation framework for process synthesis. AIChE Journal 42 (4), 1010-1032.

E. N. Pistikopoulos, N. A. Diangelakis, R. Oberdieck, M. M. Papathanasiou, I. Nascu, M. Sun, 2015. Paroc-an integrated framework and software platform for the optimisation and advanced model-based control of process systems. Chemical Engineering Science 136, 115-138.

Y. Tian, S. E. Demirel, M. F. Hasan, E. N. Pistikopoulos, 2018. An overview of process systems engineering approaches for process intensification: State of the art. Chemical Engineering and Processing-Process Intensification 133, 160-210.

A. K. Tula, D. K. Babi, J. Bottlaender, M. R. Eden, R. Gani, 2017. A computer-aided software-tool for sustainable process synthesis-intensification. Computers andamp; Chemical Engineering 105, 74-95.
S. Vedant, M. R. Atencio, Y. Tian, V. Meduri, E. N. Pistikopoulos, 2021. Towards a software prototype for synthesis of operable process intensification systems. In: Computer Aided Chemical Engineering. Vol. 50. Elsevier, pp.767-772.

732u Wirelessly Detect COVID Viruses, Bacteria, and Cells on Graphene Flakes-Improved Nano-Biosensor

Ryan Tian, Yang Tian, Ruqaiza Muhyudin, Priyangi Malaviarachchi, Roger Pechous, Xuming Zhang, John Hemmerling, Amy McCleney, Michael A. Miller, Linh Le, Steve Kaufman, Phil van Wormer, Brad Larschan, Erika Wojack, Patricia Gillenwater

In COVID-19 diagnosis, a major challenge is to develop a new methodology and a new technology that can enable us to detect live coronavirus particles simply in real-time at ultralow-cost for effectively mitigating the pandemic. As only live coronavirus is responsible for transmission and spread of disease, all ELISA- and PCR-based diagnoses cannot distinguish whether the virus is live (i.e. infectious) and whether the infected person needs quarantine, or not. To meet this long-unmet challenge, the UARK-UAMS labs have patented a new palm-sized gadget that uses reduced graphene oxide (rGO) to wirelessly detect coronaviruses and differentiate the dead virus particles from the live. Here we present new data from improving the COVID sensor using Avadain's and SWRI's high-quality, micron-sized graphene (GA) flakes (each with about three layers of carbon atoms) to replace the rGO flakes under NIST's and RAPID's support. Per our preliminary data, the COVID sensor's sensitivity, reproducibility, reliability, and manufacturing viability can be much improved using graphene flakes. On this basis, we are trying to deliver a well-optimized sensor asap that's mass-production viable for thus-detecting live viruses, bacteria, cancer cells, T- and B-cells, stem cells, etc. to potentially change games in both basic and applied biosciences. The new nanosynthesis knowledge learned from this work can potentially benefit Flextrapower's task for improving the COVID facemask.

732v Scale-up of Continuous Manufacturing and Productization of Graphene for Advanced Respirator and Biosensor Applications

John Hemmerling, Amy McCleney, Michael A. Miller, Phil van Wormer, Brad Larschan, Linh Le, Steve Kaufman, Ryan Tian, Patricia Gillenwater

Pristine, non-oxidized graphene has vast potential in numerous applications including advanced respirators and biosensors which can mitigate the devastation of future respiratory pandemic outbreaks. Specifically, higher quality respirators with better breathability to protect wearers and easy-to-use biosensors that reliably detect different types of pathogens will be critical during any future coronavirus or other outbreaks. Graphene-enhanced filters and sensors offer the promise to deliver these new tools. However, large-scale production for such strict quality control graphene has remained a challenge, and the limited supply of graphene is costly and dominated by overseas product of questionable or unacceptable quality. There is a critical need for large, thin and nearly defect free (LTDF) graphene flakes manufactured domestically at a price point that allows for democratization of this advanced material. The focus of this work is on scaling up a patented electrochemical exfoliation and post-treatment process to convert commercial-grade graphite to fully reduced graphene flakes of high structural purity, and testing the product in two critical coronavirus applications: affordable, high performance, breathable N98+ graphene-based filtration masks and low cost, accurate graphene-based biosensors to detect live coronavirus in situ. Southwest Research Institute (SwRI) has tested, validated, and optimized a bench top electrochemical reactor for small-scale (~200 mg) batch production of pristine LTDF graphene flakes. SwRI has performed quality control and quality assurance which includes characterization techniques such as powder X-ray diffractometry (PXRD), Raman spectroscopic microanalysis, Fourier-transform infrared (FTIR) spectroscopy, and scanning electron microscopy coupled with energy dispersive X-ray spectroscopy (SEM-EDS). These techniques confirm that the graphene is thin (< 5 atomic layers), non-oxidized, and nearly defect-free with large flake area (avg. 55 μm2). Aggregation of graphene flakes is a challenge, and preliminary efforts to suspend the graphene flakes with various solvents has resulted in > 40% of the graphene flake product being dispersed for weeks without settling or aggregating. Based on lessons learned from the small-scale reactor, SwRI is currently designing a medium-scale electrochemical reactor with 10x capacity to produce up to 2 grams of graphene per batch. At this scale, sufficient graphene can be provided to Flextrapower, Inc to further develop their advanced filtering and respirators, and to University of Arkansas to test with their biosensors. The ultimate goal of the project is to model, simulate, design, and construct a pilot-scale, continuous-flow reactor for large-scale synthesis, isolation, and purification of defect-free graphene flakes.

732z Accelerated Modular Process Development for Domestic Manufacturing of Critical Pharmaceutical Precursors

Nima Yazdanpanah, Cheryl Teich, John Dever, Mike Burgess, Robert Nunley, Ignasi Palou-Rivera, Patricia Gillenwater

'The pandemic lockdowns and disruptions imposed pressure on manufacturing flow and supply chain of key industrial materials and chemical reagents required for manufacturing strategic products, such as active pharmaceutical ingredients. Hence, the role of domestic manufacturing capabilities for key chemical commodities and reagents for critical products became the center of attention for the American industries and government.

The RAPID Manufacturing Institute, with its partners, was awarded a project by the U.S. Dept. of Commerce to develop a testbed for domestic manufacturing of critical pharmaceutical ingredients. The planned test bed program presents an opportunity to accelerate chemical process development with modular processing and demonstrate its potential for rapid deployment of distributed manufacturing. RAPID RACER AMPD will also deliver a process development methodology to support the design and operation of modular facilities.

We will present the activities for:

1- Current state of process development, and process development methodology to support the design and operation of modular facilities.

2- Molecule candidates selection process and final proposed products.

3- Critical chemistry knowledge and capability development for fluorination, nitration, and chlorination.

4- Modular process design, advanced manufacturing, and converting common batch processes to high throughput and high yield continuous processes.

5- Testbed development, equipment and process design, and pilot plant fabrication.'

732ac Water-Mellon: A Juicy Mini-Reactor System for Economical and Modular Water Filtration

Claire Su, Yerim Lee, Gabriel Mendez-Sanders, Lance Miller, Alayna Mikush, Devina Jain, Shivank Joshi, Justin Croyle

Our cube is designed to not only remove contaminants from water but also for sustainability. It uses two primary methods to clean and disinfect the water. First, the water flows through physical filters, including an activated carbon filter. Then, the water is treated with an aqueous sodium hypochlorite solution to treat bacteria and viruses.

Activated carbon's surface characteristics allow it to attract organic and other non-polar compounds. It has a large inner surface area with micro-pores that allows it to have high adsorption capacity for dissolved organic matter.

To address microorganisms - especially viruses - in drinking water, our process utilizes sodium hypochlorite (NaOCl) as a chemical disinfectant. When dissolved in water, sodium hypochlorite breaks down into $Na+$ and $OCl-$, which reacts with water to form HOCl. Both HOCl and $OCl-$ kill bacteria and viruses via oxidative protein unfolding, denaturing microorganisms' proteins, which kills/damages them.

AUTHOR INDEX